FROM INTELLIGENT SYSTEMS PRESS

PROCESS INTELLIGENCE PLAYBOOK

The Category-Defining Enterprise Operating Framework

Aligning Agile, DevOps, AI, Engineering, Governance, Product Delivery, and Executive Strategy

By V. Quan Lee & J. Kay

Intelligent Systems Press

INTELLIGENT
SYSTEMS PRESS
IDEAS. SYSTEMS. INTELLIGENT IMPACT.

COPYRIGHT

Published by Intelligent Systems Press
Intelligent Systems Press is an independent publishing imprint focused on system design, process intelligence, and modern delivery optimization.

ISBN: 979-8-9958725-2-8
First Edition

Editing by: Brian Loveless
Web Content Design by: Erik Victor Jones
Content Editing and Layout by: Emma Victoria Lee
For permissions, inquiries, or bulk orders, contact:
Customerservice@intelligentsystemspress.com
www.IntelligentSystemsPress.com

TABLE OF CONTENTS

INTRODUCTION

The Process Intelligence Playbook is not a theoretical companion to Agile, DevOps, AI transformation, or enterprise modernization.

It is the operational execution layer designed to synchronize enterprise strategy, governance systems, engineering reliability, AI oversight, delivery operations, product alignment, and organizational intelligence into one connected operating model.

Traditional enterprise transformation efforts often fail because organizations optimize individual functions while ignoring cross-functional synchronization.

Engineering accelerates. Governance fragments. Product priorities shift. AI systems scale. Operational visibility deteriorates.

The result is organizational acceleration without enterprise alignment.

The purpose of this Playbook is to operationalize Process Intelligence across the enterprise through:

- Operational telemetry systems
- Predictive governance models
- Delivery intelligence frameworks
- Stakeholder synchronization systems
- AI oversight structures
- Maturity scoring models
- Enterprise diagnostics
- Execution stabilization systems
- Transformation orchestration models

This Playbook is intentionally structured as an enterprise operating manual, designed to support:

- Executive Leadership
- Enterprise Strategists
- Product Organizations
- Engineering Leaders
- QA Organizations
- DevOps & Platform Teams
- Agile Delivery Teams
- Governance Leaders
- AI Transformation Programs
- Enterprise Modernization Initiatives

◆ ◆ ◆

The organizations that dominate the next decade will not merely deploy faster. They will operate more intelligently.

◆ ◆ ◆

PART I

Enterprise Alignment Diagnostics

Identifying Hidden Organizational Instability

The Enterprise Alignment Crisis

Modern organizations are no longer constrained primarily by technical capability. They are constrained by synchronization failure.

Most enterprises possess modern cloud platforms, Agile delivery teams, DevOps pipelines, automation tooling, AI-assisted systems, and digital reporting platforms. Yet despite increased technological sophistication, operational instability continues to grow.

Because operational acceleration has outpaced organizational alignment.

Enterprises deploy faster while:

- Governance visibility decreases
- Communication fragmentation increases
- Operational trust deteriorates
- Delivery predictability weakens
- Stakeholder alignment erodes

The problem is no longer merely delivery speed. The problem is enterprise synchronization.

Enterprise Failure Pattern — The Visibility Collapse

A Fortune 500 organization accelerated Agile modernization across 40 product teams while simultaneously implementing AI-assisted reporting systems and DevOps automation. Delivery throughput improved by 31%.

However: executive trust declined, governance instability increased, operational visibility fragmented, release rollback frequency grew, and escalation meetings doubled.

Engineering organizations optimized local execution. Product organizations optimized customer velocity. Governance organizations optimized compliance enforcement. None of the systems shared synchronized operational telemetry.

The organization accelerated into operational fragmentation. This is the modern enterprise alignment crisis.

Diagram Focus:

- Visibility → Alignment → Optimization loop
- Enterprise synchronization cycle
- Continuous intelligence feedback system
- Executive telemetry integration
- Adaptive operational improvement

Recommended Visual Components:

- Enterprise telemetry pathways
- AI-assisted optimization loops
- Governance synchronization layer
- Stakeholder alignment streams
- Continuous operational intelligence arrows

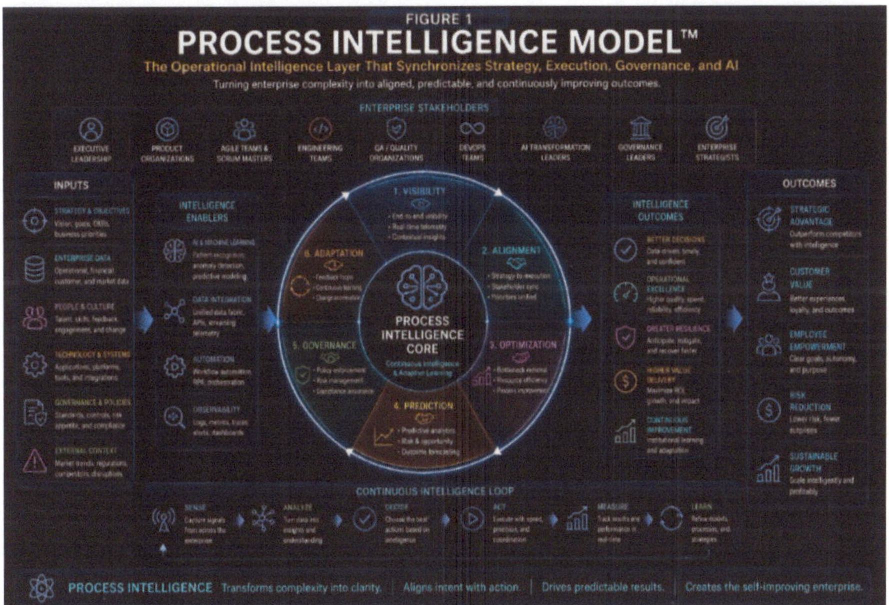

OPERATIONAL OBJECTIVE

Establish an operational intelligence layer capable of synchronizing enterprise visibility, alignment, optimization, and continuous adaptation.

Core Operational Principles

Visibility

See how work actually flows across enterprise systems.

Alignment

Synchronize stakeholders, governance systems, engineering operations, and delivery objectives.

Optimization

Continuously improve enterprise execution using operational telemetry and predictive intelligence.

Operational Warning Indicators

Organizations experiencing alignment instability commonly display:

- Increased sprint rollover
- Duplicate governance reporting
- Escalating dependency conflicts
- Release rollback growth
- Communication latency
- Unclear operational ownership
- Meeting saturation
- Executive reporting inconsistencies
- Unstable prioritization patterns
- Delayed decision cycles

These indicators represent organizational drift.

Executive Operational Actions

Executive Leadership

- Establish weekly operational telemetry reviews
- Govern transformation through measurable intelligence systems
- Align funding decisions to delivery confidence metrics
- Introduce enterprise drift monitoring systems
- Reduce governance fragmentation across departments

Product Organizations

- Prioritize work using enterprise impact scoring
- Balance customer urgency against operational stability
- Integrate dependency telemetry into roadmap planning
- Synchronize delivery objectives with platform capacity

Engineering Organizations

- Monitor architectural drift continuously
- Measure reliability degradation patterns

- Integrate deployment telemetry into executive reporting
- Track technical debt exposure quarterly

QA Organizations

- Introduce predictive defect analytics
- Implement release confidence scoring
- Establish intelligent testing visibility systems
- Integrate AI-assisted validation strategies

DevOps & Platform Teams

- Measure deployment confidence continuously
- Monitor environment instability trends
- Govern automation through operational safeguards
- Establish telemetry-aware release systems

AI Governance Leaders

- Introduce explainability checkpoints
- Govern AI escalation pathways
- Monitor AI operational drift
- Align AI acceleration with governance policy

Diagram Focus:

- Executive operational KPI visibility
- Enterprise alignment scoring
- Governance telemetry systems
- Delivery confidence analytics

Recommended Visual Components:

- KPI scorecards
- Operational trend analytics
- Alignment scoring gauges
- Predictive forecasting indicators
- Governance heat maps
- Transformation stability tracking

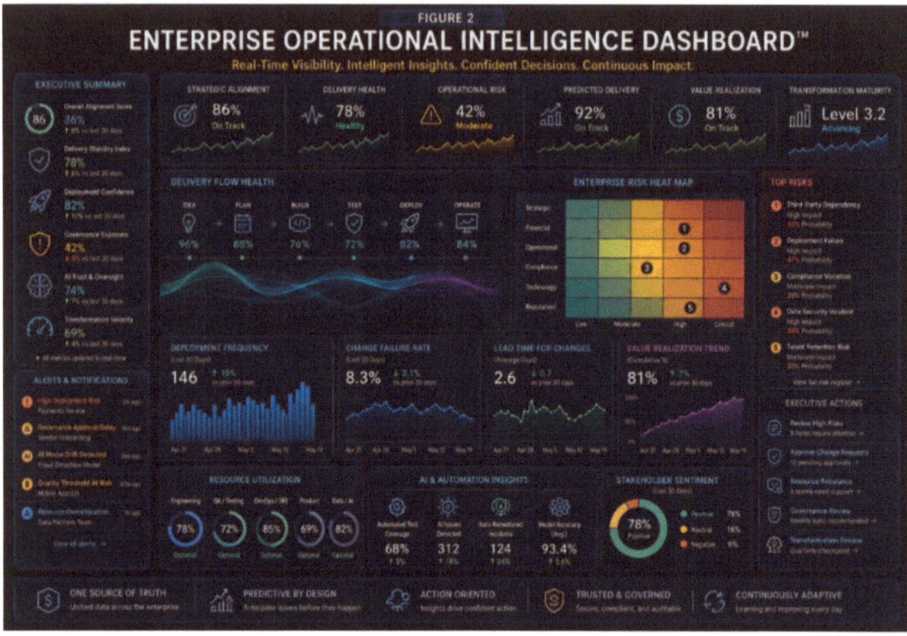

FIGURE 2
ENTERPRISE OPERATIONAL INTELLIGENCE DASHBOARD™

Enterprise Metrics

Metric	Operational Meaning
Enterprise Drift Score	Measures organizational misalignment
Delivery Stability Index	Measures release consistency
Governance Exposure Rating	Measures operational oversight risk
Synchronization Index	Measures cross-functional cohesion
Transformation Confidence Score	Measures enterprise modernization stability

PART II

Operational Intelligence Systems

Synchronizing Enterprise Delivery

The Enterprise Operating System

Traditional enterprises operate through disconnected functional structures. Product teams plan independently. Engineering teams optimize independently. Governance teams audit independently. Executives measure independently.

Disconnected optimization creates invisible enterprise instability.

Process Intelligence establishes a synchronized operational ecosystem connecting executive strategy, delivery execution, governance systems, engineering telemetry, AI oversight, operational intelligence, and organizational adaptation.

The future enterprise is not merely faster. It is synchronized.

Diagram Focus:

- Executive operational visibility
- Strategic governance intelligence
- Predictive delivery oversight
- Enterprise telemetry command system
- Risk escalation visibility

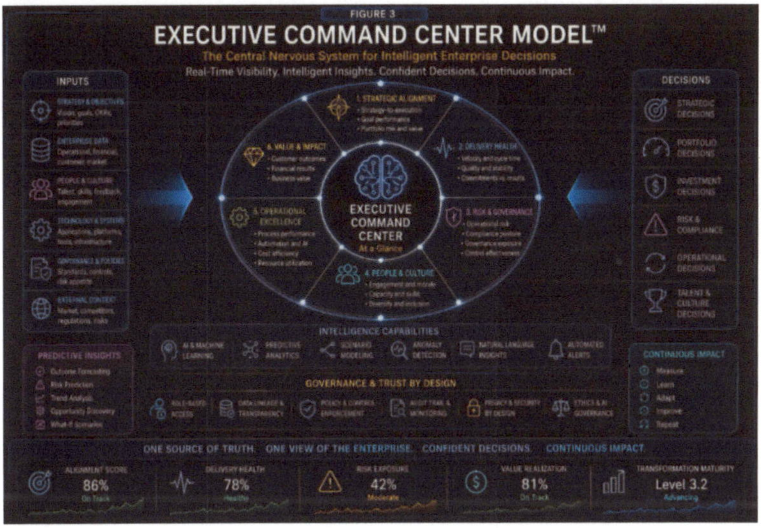

Recommended Visual Components:

- Central executive intelligence hub
- Delivery health indicators
- Governance exposure layer
- AI operational insights
- Enterprise KPI visualization
- Predictive decision pathways

> **OPERATIONAL OBJECTIVE**
> Create a centralized operational intelligence system capable of providing leadership with strategic visibility, predictive delivery intelligence, governance exposure awareness, transformation stability indicators, and operational confidence forecasting.

Executive Visibility Crisis

Many organizations believe dashboards create visibility. They do not.

Most dashboards measure isolated metrics rather than synchronized operational intelligence. Executives often see delivery throughput, sprint completion, deployment frequency, and financial reporting — but fail to see hidden dependency instability, governance latency, stakeholder fragmentation, architectural drift, and quality degradation trends.

This creates executive blindness disguised as operational transparency.

Executive Command Actions

Establish Enterprise Telemetry Reviews

Executives should review synchronized delivery telemetry weekly rather than relying solely on quarterly reporting structures.

Governance Through Predictive Intelligence

Operational forecasting must replace reactive escalation management.

Centralize Operational Visibility

Create unified visibility across engineering, product, governance, AI systems, and platform operations.

Introduce Enterprise Drift Detection

Monitor hidden instability patterns before operational disruption impacts customers.

Enterprise Scenario — The Reporting Illusion

A global enterprise implemented advanced reporting dashboards across delivery organizations. Leadership initially believed visibility had improved.

However, every department measured success differently. Engineering reported deployment velocity. Product organizations reported roadmap progress. Governance teams reported compliance status. Executives reported transformation milestones.

No shared operational intelligence layer existed. The dashboards increased reporting volume while decreasing enterprise understanding.

Process Intelligence resolved the fragmentation by introducing synchronized telemetry systems, unified governance metrics, operational confidence scoring, and predictive delivery intelligence.

Recommended Visual Components:

- Executive leadership node
- Product + engineering + QA + DevOps alignment streams
- Governance oversight integration
- Shared operational telemetry hub
- Cross-functional synchronization loops

OPERATIONAL OBJECTIVE

Synchronize enterprise stakeholders around shared operational intelligence systems.

Organizational Misalignment Patterns

Cross-functional organizations rarely fail because teams lack talent. They fail because systems optimize independently:

- Product teams prioritizing speed while engineering prioritizes stability
- Governance organizations slowing delivery through fragmented approval systems
- DevOps teams accelerating deployment without operational oversight integration
- AI transformation programs scaling faster than governance maturity

Misalignment compounds over time. This creates synchronization debt.

Stakeholder Synchronization Actions

Executive Leadership

- Establish enterprise decision alignment reviews
- Standardize transformation reporting models
- Align operational KPIs across departments

Product Organizations

- Integrate operational telemetry into roadmap planning

- Measure customer value against delivery sustainability
- Prioritize based on enterprise impact scoring

Engineering Organizations

- Synchronize architectural standards enterprise-wide
- Measure reliability trends continuously
- Integrate platform observability into governance systems

QA Organizations

- Centralize release confidence reporting
- Measure quality degradation patterns proactively
- Implement intelligent validation workflows

Governance Organizations

- Simplify escalation pathways
- Eliminate duplicate approval systems
- Align oversight systems with delivery telemetry

Delivery Synchronization Metrics

Metric	Meaning
Stakeholder Alignment Score	Measures cross-functional cohesion
Escalation Frequency Rating	Measures governance instability
Dependency Conflict Index	Measures delivery coordination risk
Communication Latency Score	Measures decision responsiveness
Operational Ownership Clarity	Measures accountability stability

Diagram Focus:

- Intelligent prioritization systems
- Enterprise value optimization
- Delivery complexity analysis
- Strategic impact forecasting

Recommended Visual Components:

- Multi-axis value scoring matrix
- Customer impact layer
- Governance risk indicators
- Dependency analysis pathways
- Operational capacity intelligence
- Prioritization heat mapping

Strategic Prioritization

OPERATIONAL OBJECTIVE
Prioritize enterprise initiatives using synchronized operational intelligence rather than isolated business urgency.

Strategic Prioritization Failure Modes

Many enterprises prioritize initiatives using political urgency, executive pressure, roadmap optimism, or isolated customer demand. This creates delivery overload, platform instability, operational exhaustion, and governance fragmentation.

The result is enterprise acceleration without sustainability.

Intelligent Prioritization Actions

Every initiative should be evaluated using:

- Customer impact
- Operational complexity
- Governance exposure
- Architectural dependency
- Delivery stability
- Organizational capacity

Introduce Delivery Capacity Intelligence

Roadmaps must align to operational capability rather than aspirational planning.

Forecast Dependency Risk

Organizations must identify hidden coordination conflicts before delivery begins.

Enterprise Scenario — The Overloaded Roadmap

A digital enterprise launched simultaneous transformation initiatives across multiple business units. Customer demand justified every initiative.

However, engineering capacity was overloaded, governance reviews slowed execution, deployment reliability deteriorated, and operational fatigue increased.

The organization optimized strategic ambition while ignoring delivery sustainability.

Process Intelligence stabilized execution by introducing enterprise capacity intelligence, dependency scoring systems, governance-aware prioritization, and operational sustainability metrics.

PART III

Delivery Governance & Execution Control

The Intelligent Delivery Layer

Diagram Focus:

- Operational instability diagnostics
- Delivery disruption visibility
- Communication drift analytics
- Friction detection telemetry

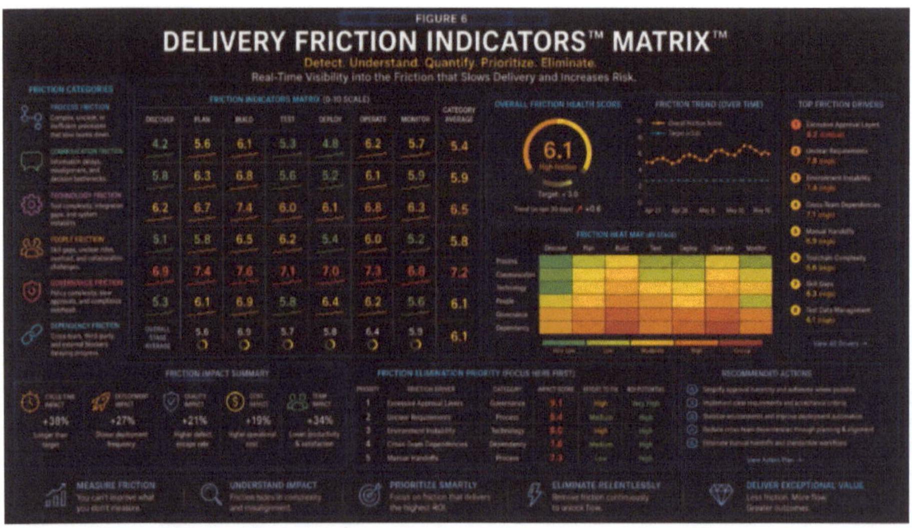

Recommended Visual Components:

- Friction scoring dashboard
- Communication instability metrics
- Governance latency indicators
- Dependency overload visualization
- Escalation trend graphs
- Operational bottleneck heat zones

OPERATIONAL OBJECTIVE
Identify hidden execution instability before organizational performance deteriorates.

Delivery Friction Diagnostics

Delivery friction rarely appears immediately. It accumulates silently through communication overload, dependency instability, governance latency, meeting saturation, unclear ownership, and fragmented escalation systems.

Organizations often normalize operational inefficiency until instability becomes systemic.

The Delivery Friction Index™

Friction Area	Operational Impact
Communication Drift	Delayed alignment
Dependency Overload	Sprint instability
Governance Latency	Slowed execution
Release Volatility	Reduced operational trust
Meeting Saturation	Cognitive overload
Ownership Ambiguity	Accountability failure

Operational Response Actions

If friction increases:

- Reduce dependency complexity
- Simplify governance pathways
- Clarify operational ownership
- Consolidate reporting systems
- Automate telemetry visibility
- Stabilize escalation structures

Enterprise Scenario — The Coordination Collapse

A large-scale modernization initiative introduced additional coordination meetings to improve alignment. The result was the opposite.

Decision velocity slowed. Execution latency increased. Stakeholders spent more time discussing delivery than executing delivery.

Meeting volume became a hidden operational tax.

Process Intelligence reduced friction by simplifying communication pathways, centralizing telemetry visibility, clarifying operational ownership, and automating escalation workflows.

Diagram Focus:

- AI-enhanced CI/CD governance
- Predictive deployment intelligence
- Quality confidence systems
- Enterprise release telemetry

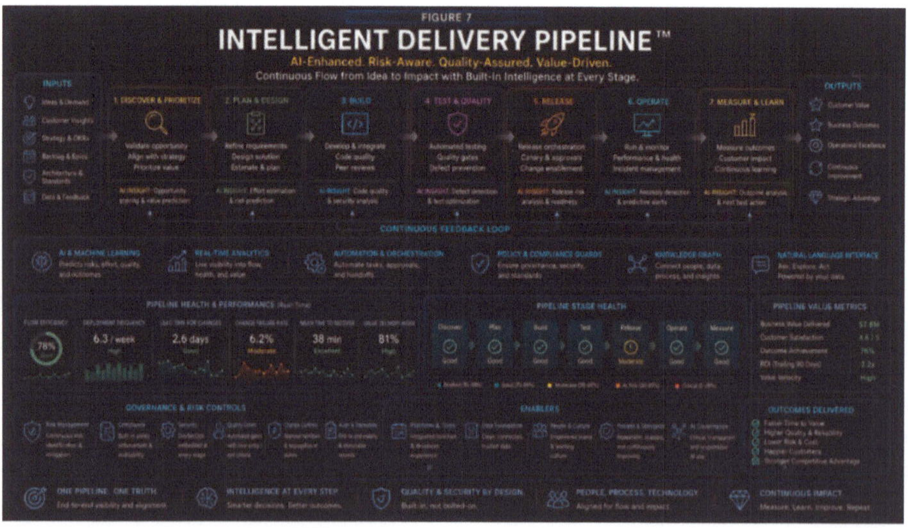

Recommended Visual Components:

- Pipeline lifecycle stages
- Automated quality gates
- Deployment risk scoring
- Rollback prediction systems

- Governance checkpoints
- Telemetry-driven release controls

Intelligent CI/CD & Deployment Governance

OPERATIONAL OBJECTIVE
Establish enterprise CI/CD systems governed through predictive operational intelligence.

Deployment Confidence Engineering

Traditional delivery pipelines measure deployment frequency, build success, and release throughput. Intelligent delivery pipelines must additionally measure:

- Rollback probability
- Governance exposure
- Quality confidence
- Operational risk
- Dependency instability
- AI-assisted deployment risk

Intelligent Delivery Actions

DevOps & Platform Teams

- Implement deployment confidence modeling
- Introduce predictive rollback forecasting
- Govern automation through operational safeguards
- Centralize release telemetry systems

QA Organizations

- Integrate predictive validation analytics
- Measure release quality confidence continuously
- Automate operational quality telemetry

Engineering Organizations

- Monitor deployment stability trends
- Forecast platform reliability degradation
- Integrate observability into executive reporting

Intelligent Delivery Metrics

Metric	Purpose
Deployment Confidence Score	Forecasts release stability
Rollback Probability Index	Predicts release failure risk
Platform Reliability Rating	Measures infrastructure stability
Quality Confidence Score	Measures validation reliability
Operational Telemetry Coverage	Measures visibility maturity

PART IV

Predictive Governance & AI Oversight

Governing Intelligent Enterprises

Diagram Focus:

- AI governance orchestration
- Human-in-the-loop oversight
- Explainability intelligence
- Operational trust systems

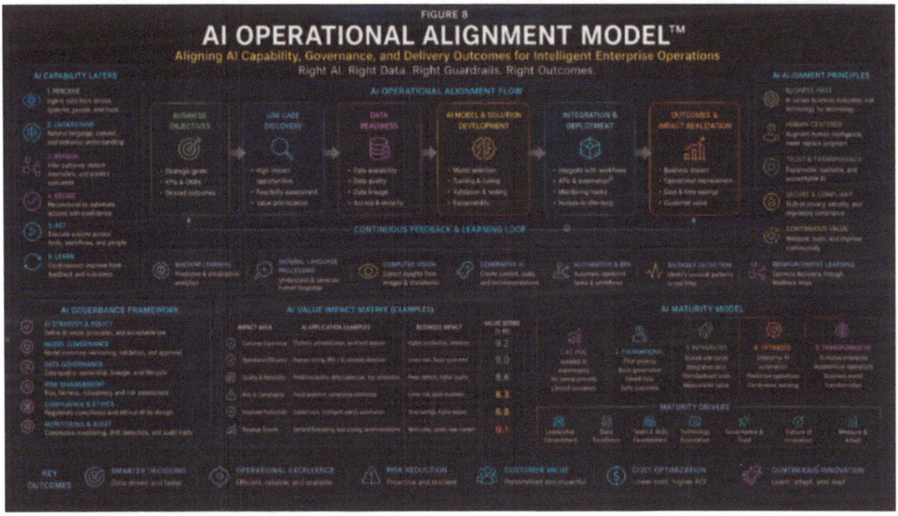

Recommended Visual Components:

- AI decision governance pathways
- Escalation control layers
- Human oversight checkpoints
- AI trust scoring systems
- Enterprise compliance alignment
- Operational intelligence synchronization

OPERATIONAL OBJECTIVE
Govern AI acceleration through synchronized operational oversight.

AI Governance Escalation Models

Most organizations focus heavily on AI capability. Far fewer focus on AI governance maturity, escalation pathways, operational safeguards, explainability controls, and enterprise accountability.

AI acceleration without synchronized governance creates invisible enterprise risk.

AI Governance Actions

Establish Human-in-the-Loop Controls

Critical enterprise decisions must retain escalation safeguards.

Introduce Explainability Reviews

AI systems must provide operational transparency.

Govern AI Operational Drift

Monitor how AI recommendations evolve over time.

Align AI With Enterprise Policy

Operational intelligence must remain synchronized with governance standards.

Enterprise Scenario — AI Acceleration Without Governance

A global enterprise accelerated AI-assisted deployment systems across multiple business units. Delivery speed improved dramatically.

However, escalation pathways were inconsistent, explainability standards varied, governance visibility fragmented, and operational trust deteriorated.

Engineering trusted automation. Governance trusted reporting. Leadership trusted dashboards. None of the systems shared synchronized operational intelligence.

Process Intelligence resolved the fragmentation through unified operational telemetry, synchronized governance reviews, AI trust checkpoints, and predictive escalation systems.

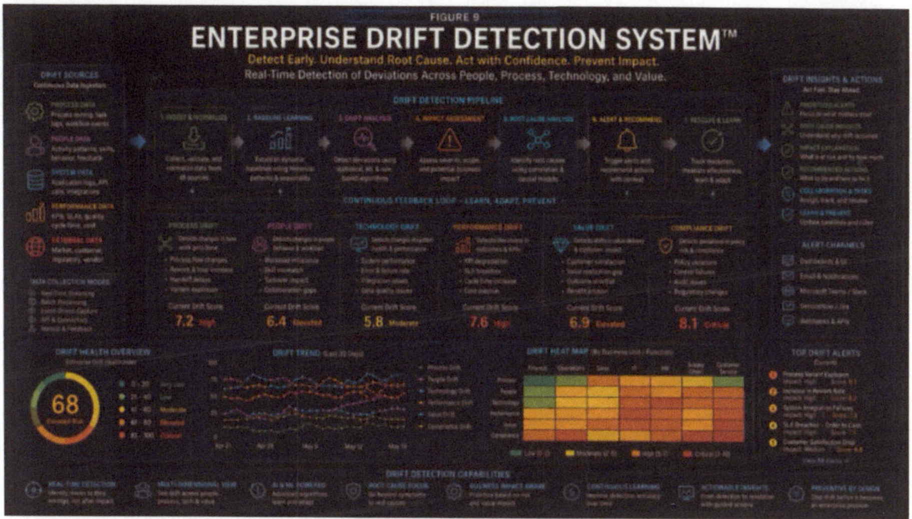

Diagram Focus:

- Predictive operational instability detection
- Organizational drift analytics
- Strategic misalignment monitoring
- Enterprise stabilization systems

Recommended Visual Components:

- Drift detection engine
- Operational risk indicators
- Governance exposure alerts
- Strategic deviation monitoring
- Predictive escalation pathways
- Enterprise stabilization feedback loops

Organizational Drift Detection

> **OPERATIONAL OBJECTIVE**
> Identify operational instability before enterprise disruption occurs.

Organizational Drift Patterns

Enterprise drift occurs when:

- Strategy and execution diverge
- Governance and delivery disconnect
- AI systems evolve faster than oversight
- Product priorities conflict with operational capacity
- Engineering reliability weakens silently

Drift rarely appears as a single catastrophic event. It emerges gradually through accumulated instability.

Drift Detection Actions

Monitor Early Warning Signals

- Escalation growth
- Dependency instability
- Release rollback increases
- Governance approval latency
- Communication fragmentation

Forecast Operational Risk

Use predictive telemetry rather than reactive reporting.

Introduce Enterprise Stabilization Reviews

Synchronize leadership, engineering, governance, and delivery systems continuously.

Predictive Governance Metrics

Metric	Purpose
Enterprise Drift Score	Measures organizational instability
Predictive Stability Index	Forecasts disruption probability
Governance Responsiveness Rating	Measures escalation efficiency
AI Trust Index	Measures explainability maturity
Operational Confidence Score	Measures execution reliability

PART V

Self-Optimizing Enterprise Operations

The Future Intelligent Enterprise

Diagram Focus:

- Future-state enterprise intelligence
- Predictive operational orchestration
- Adaptive governance systems
- AI-assisted forecasting

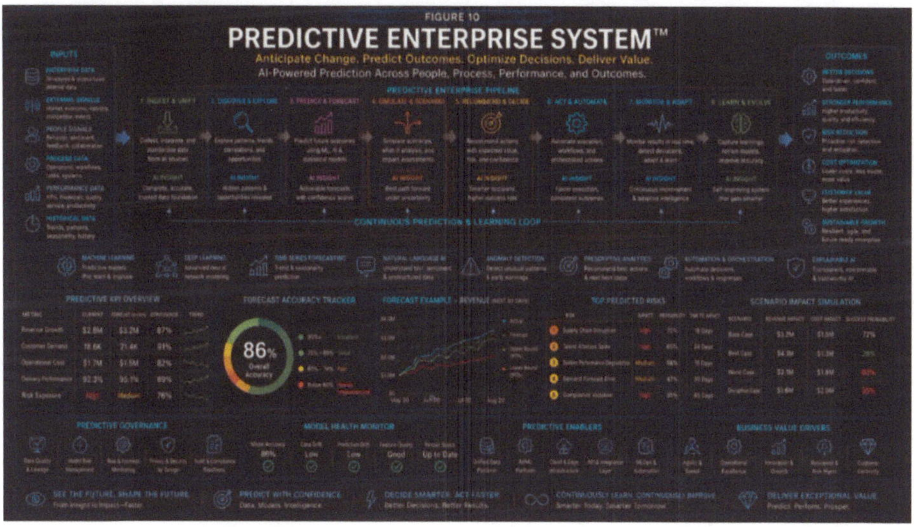

Recommended Visual Components:

- Predictive analytics engine
- Enterprise forecasting dashboards
- Delivery intelligence flows
- AI governance synchronization
- Operational adaptation loops
- Continuous learning systems

OPERATIONAL OBJECTIVE

Create enterprises capable of forecasting operational instability before disruption impacts customers.

Predictive Decision Architectures

Future enterprises will operate through predictive telemetry, AI-assisted governance, synchronized delivery intelligence, operational forecasting, and continuous enterprise learning. **Organizations will transition from reactive management into predictive operational orchestration.**

Predictive Enterprise Actions

Executives:

- Govern through operational forecasting
- Prioritize investments using delivery intelligence
- Align strategy with predictive telemetry

Product Organizations:

- Forecast customer impact proactively
- Integrate operational intelligence into roadmap planning

Engineering Organizations:

- Predict reliability degradation before customer impact
- Forecast architectural instability trends

Governance Leaders:

- Anticipate operational exposure before escalation occurs
- Synchronize oversight with delivery intelligence

Diagram Focus:

- Autonomous enterprise intelligence
- Self-correcting operational systems
- Adaptive organizational orchestration
- Continuous optimization ecosystem

Recommended Visual Components:

- Central enterprise intelligence core
- Continuous learning loops
- AI-assisted optimization pathways
- Adaptive governance architecture
- Self-healing delivery systems
- Enterprise-wide synchronization flows

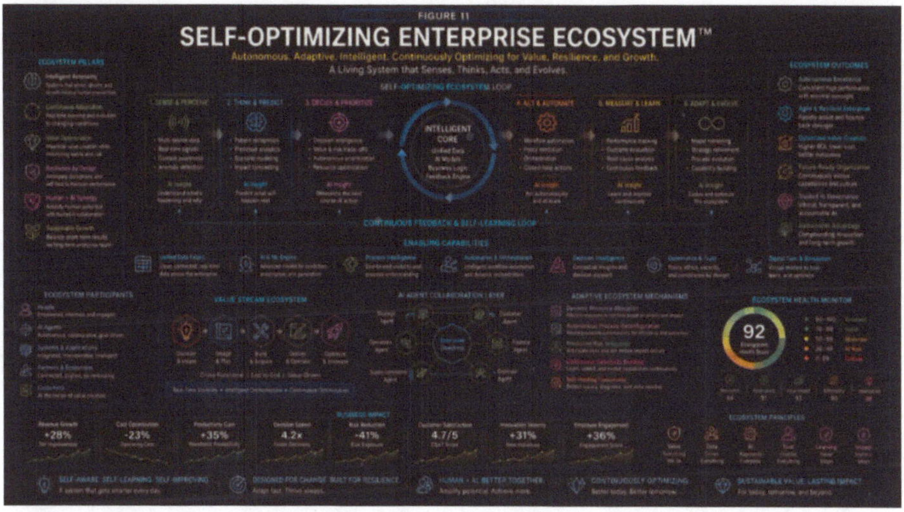

Operational Objective

Establish continuously adaptive enterprises capable of synchronized intelligence-driven optimization.

The Self-Optimizing Enterprise

The future enterprise will not operate through static organizational structures. It will function as an adaptive intelligence ecosystem capable of:

- Continuous learning
- Synchronized governance
- AI-assisted orchestration
- Predictive delivery optimization
- Operational self-correction

Diagram Focus:

- Enterprise maturity progression
- Reactive → Self-Optimizing transformation path
- Governance evolution
- Predictive intelligence advancement

Recommended Visual Components:

- 5-level maturity staircase
- Capability progression layers
- Governance maturity curve
- Operational intelligence scaling model
- Enterprise evolution pathways

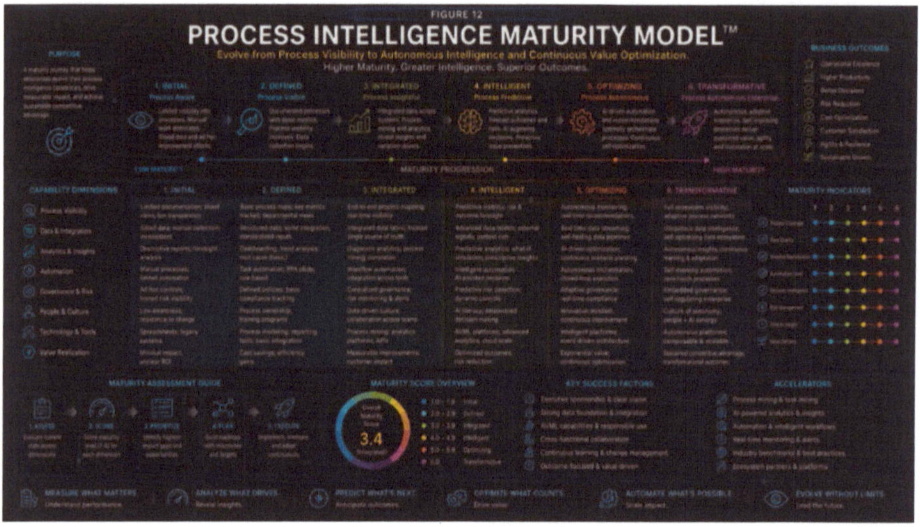

This represents the transition from fragmented execution into intelligent enterprise synchronization.

The Process Intelligence Maturity Model™

Level	Maturity Stage	Characteristics
Level 1	Reactive Enterprise	Siloed operations, fragmented governance, unstable delivery systems, manual escalation pathways
Level 2	Visible Enterprise	Dashboards implemented, partial operational transparency, inconsistent synchronization, fragmented telemetry
Level 3	Coordinated Enterprise	Synchronized delivery systems, integrated governance visibility, enterprise telemetry alignment, improved delivery predictability
Level 4	Predictive Enterprise	Operational forecasting, intelligent prioritization, drift detection systems, predictive governance controls
Level 5	Self-Optimizing Enterprise	Adaptive governance, AI-assisted orchestration, autonomous operational intelligence, continuously evolving enterprise ecosystems

Enterprise Alignment Readiness Assessment

Operational Readiness Checklist

- Is governance integrated into delivery telemetry?
- Are AI systems governed through escalation pathways?
- Are product priorities synchronized with operational capacity?
- Can leadership identify operational instability in real time?
- Are deployment risks forecasted before release?
- Are quality systems integrated into enterprise intelligence?
- Is operational ownership clearly defined?
- Are dependency conflicts measured proactively?

Delivery Stability Assessment

Capability	Score (1–5)
Governance Synchronization	
Delivery Predictability	
AI Oversight Maturity	
Cross-Functional Visibility	
Quality Intelligence	
Operational Telemetry	
Escalation Responsiveness	
Enterprise Alignment	

Operational Decision Matrix

Indicator	Operational Meaning	Executive Response
Sprint rollover increase	Planning instability	Reduce dependency overload
Rollback growth	Quality confidence decline	Trigger governance review
Meeting volume spikes	Alignment fragmentation	Simplify communication pathways
Escalation growth	Governance instability	Centralize operational ownership
Telemetry inconsistency	Visibility fragmentation	Standardize reporting systems

Signature Process Intelligence Vocabulary

Term	Definition
Organizational Drift	Hidden misalignment between enterprise strategy and operational execution
Governance Friction	Operational slowdown caused by disconnected oversight systems
Delivery Telemetry	Real-time operational intelligence signals
Synchronization Debt	Accumulated cross-functional misalignment
Predictive Governance	Forward-looking operational risk management
Execution Visibility	Measurable transparency across enterprise delivery systems
Adaptive Enterprise Layer	The intelligence system coordinating enterprise transformation

PART VI

Enterprise Transformation Failure Patterns

Why Modern Enterprise Transformations Collapse

The Illusion of Transformation Progress

Many enterprises mistake operational activity for transformation maturity. Organizations deploy Agile frameworks, DevOps tooling, AI-assisted systems, cloud modernization platforms, enterprise dashboards, and automation pipelines — yet transformation instability continues to increase.

Because transformation acceleration without operational synchronization creates systemic fragmentation.

Enterprise Anti-Patterns

Anti-Pattern	Description
Dashboard Theater	Organizations generate reporting visibility without operational intelligence
Governance Gridlock	Approval systems multiply faster than delivery coordination improves
Automation Illusion	Automation accelerates unstable systems faster
Agile Fragmentation	Teams iterate independently without enterprise synchronization
AI Acceleration Trap	AI systems scale operational decisions faster than governance systems can respond
Meeting Saturation	Organizations replace execution with excessive coordination overhead

Why Enterprise Transformations Fail

Agile Without Governance

Delivery accelerates while enterprise visibility deteriorates.

DevOps Without Operational Intelligence

Automation scales instability across environments.

AI Without Oversight

Decision acceleration outpaces accountability structures.

Metrics Without Synchronization

Departments optimize locally while enterprise systems fail globally.

Transformation Without Ownership

No operational layer exists to coordinate enterprise execution.

The Hidden Cost of Organizational Drift

Organizational drift creates measurable enterprise damage through:

- Duplicate work
- Governance delays
- Stakeholder fatigue
- Delivery instability
- Operational mistrust
- Transformation slowdown
- Escalation overload
- Communication fragmentation
- Decision paralysis
- Talent burnout

The cost of operational misalignment compounds silently over time.

Decision Velocity Metrics™

Metric	Operational Meaning
Escalation Resolution Time	Measures responsiveness efficiency
Governance Approval Cycle Time	Measures decision friction
Stakeholder Response Latency	Measures coordination delays
Release Decision Confidence	Measures operational trust
Executive Alignment Velocity	Measures strategic synchronization

PART VII

Quality Intelligence Operations

Quality as an Enterprise Intelligence System

Quality Intelligence as a Strategic Capability

Most enterprises treat quality assurance as a downstream validation activity. Intelligent enterprises treat quality as a predictive operational intelligence system.

Quality telemetry provides some of the earliest indicators of operational instability, release degradation, dependency conflict, architectural drift, governance weakness, and AI reliability risk.

Quality intelligence therefore becomes an enterprise stabilization mechanism.

Diagram Focus:

- Predictive quality engineering
- Release confidence systems
- AI-assisted validation orchestration
- Quality telemetry intelligence
- Enterprise quality governance

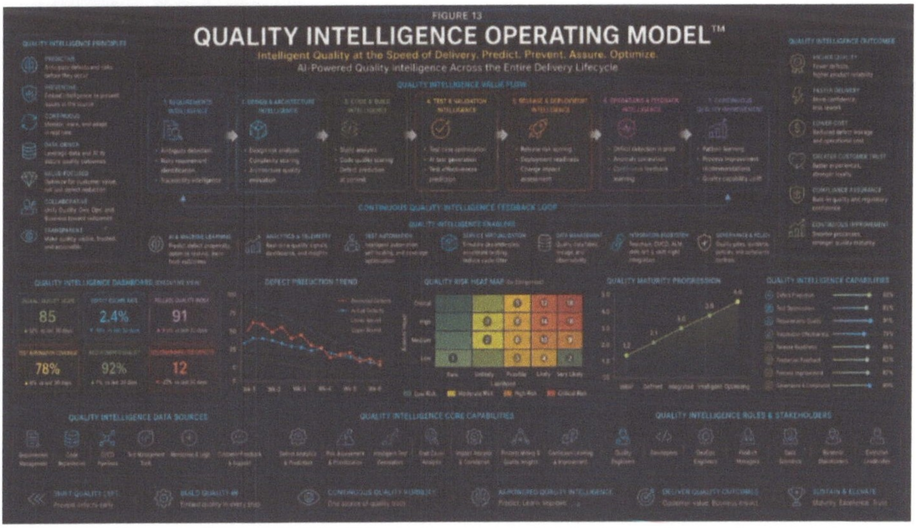

Recommended Visual Components:

- Validation intelligence engine
- Predictive defect analytics
- Automated quality telemetry
- Release confidence scoring
- Governance-integrated quality gates
- Continuous quality learning loops

Quality Intelligence Operations

Predictive Defect Forecasting

Use telemetry systems to identify quality degradation patterns before release instability occurs.

Release Confidence Engineering

Measure operational trustworthiness continuously rather than relying solely on pass/fail validation.

AI-Assisted Validation Systems

Integrate intelligent testing orchestration capable of adapting validation depth based on operational risk.

Quality Drift Detection

Monitor degradation trends across:

- Automation reliability
- Deployment consistency
- Test coverage health
- Validation confidence
- Operational resilience

Enterprise Scenario — The False Green Release

A large enterprise maintained consistently successful deployment pipelines with high automated test pass rates.

However, customer-reported defects increased, operational instability escalated, rollback frequency grew, and stakeholder confidence deteriorated.

The organization optimized automation completion while ignoring operational quality intelligence.

Process Intelligence resolved the instability through release confidence scoring, operational quality telemetry, predictive defect forecasting, and intelligent validation escalation.

PART VIII

Enterprise Stabilization & Recovery

Restoring Operational Alignment

Enterprise Stabilization Protocols™

Transformation instability is inevitable. Intelligent enterprises are distinguished not by the absence of disruption, but by the ability to recover operational alignment rapidly.

Stabilization Protocol — Delivery Recovery

When delivery predictability deteriorates:

- Reduce dependency complexity
- Centralize operational telemetry
- Stabilize governance pathways
- Re-establish ownership accountability
- Introduce release confidence monitoring

Stabilization Protocol — Governance Recovery

When governance fragmentation emerges:

- Eliminate duplicate approval systems
- Simplify escalation pathways
- Synchronize reporting standards
- Centralize operational visibility
- Re-align executive telemetry systéms

Stabilization Protocol — AI Oversight Recovery

When AI trust deteriorates:

- Reintroduce human escalation checkpoints
- Expand explainability visibility
- Audit operational decision pathways
- Increase governance synchronization reviews
- Stabilize enterprise AI accountability

PART IX

Human Synchronization Systems

The Human Layer of Enterprise Intelligence

Human Synchronization in Intelligent Enterprises

Enterprise transformation is not solely a technical challenge. It is a human synchronization challenge.

Organizations frequently underestimate the operational impact of:

- Cognitive overload
- Transformation fatigue
- Communication fragmentation
- Decision exhaustion
- Operational mistrust
- Stakeholder burnout
- Governance confusion

These human-system failures often become the hidden drivers of enterprise instability.

Human-Centered Operational Intelligence

Intelligent enterprises must continuously monitor:

- Communication clarity
- Escalation fatigue
- Coordination overload
- Organizational trust
- Stakeholder alignment health
- Operational cognitive burden

Operational intelligence therefore includes both system telemetry and human synchronization telemetry.

PART X

Process Intelligence Activation Roadmap

Operationalizing the Enterprise Transformation

90-Day Process Intelligence Activation Plan™

Phase	Name	Objectives
Phase 1	Visibility	Centralize reporting systems • Establish enterprise telemetry • Identify operational drift • Standardize governance visibility
Phase 2	Alignment	Synchronize stakeholders • Stabilize governance pathways • Reduce dependency fragmentation • Establish executive telemetry reviews
Phase 3	Prediction	Introduce forecasting systems • Implement intelligence metrics • Establish drift detection • Monitor operational instability proactively
Phase 4	Optimization	Automate orchestration systems • Implement adaptive governance • Optimize enterprise synchronization • Evolve into predictive enterprise operations

PART XI

The Rise of the Intelligent Enterprise

The Future Operating Model

The Enterprise of the Future

The future enterprise will not merely be digital. It will be operationally intelligent.

Future organizations will operate through:

- Predictive governance
- Adaptive orchestration
- Synchronized delivery intelligence
- AI-assisted operational systems
- Continuous enterprise learning
- Self-healing operational ecosystems

Operational intelligence will become as foundational to enterprises as financial intelligence.

The enterprises that dominate the next decade will not simply automate faster. They will:

- Align continuously
- Govern predictively
- Adapt operationally
- Synchronize intelligently
- Optimize sustainably
- Stabilize dynamically

The future enterprise is not merely automated. It is operationally intelligent.

FINAL STRATEGIC POSITIONING

Process Intelligence is not merely another Agile framework, DevOps methodology, AI implementation model, or process improvement philosophy.

It is a category-defining enterprise operating framework designed to synchronize:

- Leadership Strategy
- Governance Intelligence
- Agile Delivery
- Engineering Reliability
- Product Alignment
- DevOps Execution
- AI Oversight

- Operational Visibility
- Predictive Intelligence
- Enterprise Adaptability

◆ ◆ ◆

The future enterprise is not faster.

It is aligned.

◆ ◆ ◆

Process Intelligence is the operational intelligence layer connecting Agile execution, DevOps delivery, AI systems, governance intelligence, engineering reliability, operational visibility, predictive analytics, and enterprise strategy into one synchronized ecosystem.

V. Quan Lee & J. Kay

Intelligent Systems Press

www.ingramcontent.com/pod-product-compliance
Lightning Source LLC
LaVergne TN
LVHW051132210926
848482LV00001B/68
* 9 7 9 8 9 9 5 8 7 2 5 2 8 *